THE CELL WORKS

MICROEXPLORERS

First edition for the United States and Canada
published exlusively 1997 by Barron's Educational Series, Inc.

Originally published in English under the title *Sensational Cells*
© Copyright USEFUL BOOKS, S.L., 1997 Barcelona, Spain.

Authors: Patrick A. Baeuerle and Norbert Landa
Illustrators: Antonio Muñoz and Roser Rius
Graphic Design: IGS – Barcelona, Spain

Address all inquiries to:
Barron's Educational Series, Inc.
250 Wireless Boulevard
Hauppauge, NY 11788

Library of Congress Catalog Card No. 97-74822

International Standard Book Number 0-7641-5052-9

Printed in Spain

987654321

THE CELL WORKS

MICROEXPLORERS

*An expedition
into the fantastic
world of cells*

*Norbert Landa and
Patrick A. Baeuerle*

BARRON'S

Welcome to the

Hello everybody, and welcome aboard our MicroMachine. I am Professor Gene, your tour guide. This is our MicroMachine, which will take us on an expedition into the fantastic world of cells.

In a few seconds, we will start the shrinking process. The MicroMachine will allow us to shrink so small that we can travel anywhere—into the body, in between the cells, and even inside of the cells. We will become several thousand times—even a hundred thousand times—smaller than we are now. We will stop at interesting and sometimes very strange spots. We will watch amazing things take place—things we normally would never even get a glimpse of!

We want to find out why you and I and elephants and trees are so different. We also want to learn what the difference is between living things, such as people, and nonliving things, such as stones and machines.

In this way, we will learn about lots of things that happen in our cells—how we are able to grow, run, or think about things. We will also learn why we need air and nutrients to keep alive.

Many things will appear strange and almost unbelievable. Those are the very things that are taking place in each one of us a million times every second.

If you have any questions along the way, don't hesitate to ask me!

We'll start in a moment.

tour!

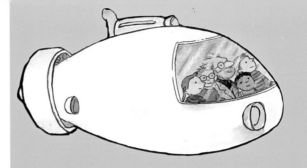

First, a little

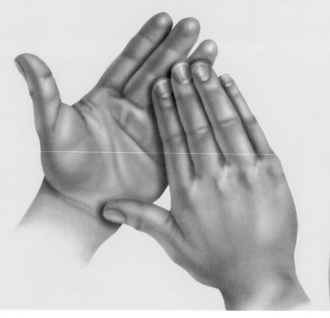

Normally, a single cell is much too small to be seen or felt. Even when we hold our hands under a magnifying glass, we can see only a solid surface of skin cells. When we rub our hands, however, thousands of dead skin cells stick together and form strange little lumps.

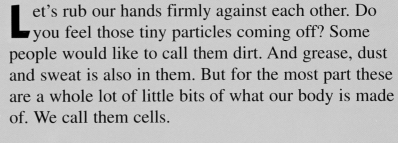

experiment

Let's rub our hands firmly against each other. Do you feel those tiny particles coming off? Some people would like to call them dirt. And grease, dust and sweat is also in them. But for the most part these are a whole lot of little bits of what our body is made of. We call them cells.

Beneath these dead skin cells that continuously flake off are living skin cells. Between and beneath the living skin cells, deep inside the fingers, are many other kinds of cells at work. Altogether, there are several hundred different kinds of cells in the human body. Each kind looks different from the others and has special tasks to fulfill.

Each of us consists of many billions of cells. They are constantly working to keep us alive. Cells grow and multiply. They exchange messages with each other. They create energy. Cells absorb nutrients from their surroundings and change them into the very substances that the cell itself is made of. Each cell follows an exact plan that perfectly coincides with the working plans in other cells. The result is a living organism. An organism is a complete living thing.

Tell us, Gene, are animals and plants also made of cells?

Yes, they all consist of cells—even bacteria, though these consist of just one single cell. All kinds of cells within all kinds of living beings work more or less like human cells. Of course, organisms differ quite a lot. Cabbage, for example, have no brains. Humans, however, have an incredible number of nerve cells that work together in a particularly complicated manner in our brain. People are the only inhabitants of our planet who are able to explore questions like what makes a living creative a living creature. So let's not miss our chance to do so!

Okay, we have now reached Shrink Stage One. We are on our way!

What happens in our *skin*?

Do you see that giant rod? In reality, it is a very fine hair on the finger. We will use it as an entryway into the skin of the finger.

Skin cells are constantly busy protecting the body throughout the three weeks that they live. They fend off dust and dirt, hazardous light, and, most importantly, bacteria and other tiny creatures trying to invade us. At some parts of the body's surface, for example on our heels, skin cells make very thick layers. At other places, however, the skin stays thin and sensitive.

Gene, why do these cells fly away?

They have accomplished their task. Every second of our lives we lose thousands of dead skin cells. At the same time, fresh cells are formed inside the skin and grow up. They replace cells that are dead and gone. Actually, all the outer cells of our body are no longer alive. This includes the top layer of our skin, our hair, our nails, and even the surface of our tongue and eyes.

Under the top layer of the skin, life is going on. We see here cells of different types. At the hair root, cells constantly divide to form fresh cells. Whey they die, the cells assemble as hair and are pushed up.

Why does it hurt, then, when we pull out a hair?

The hair's root is still connected with the fine extensions of the nerve cells. Nerve cells even register when a hair is moved. Nerve cells are what make the skin so sensitive. They allow us to feel pressure, heat, and cold. Nerve cells also give command signals to our muscles.

What is that little sack down at the root of the hair?

That is the sebaceous gland. Its cells produce a kind of grease that keeps the hair smooth, and with time, makes it greasy. That curled-up tubing nearby is one of the sweat glands, which can spread little drops of salt water onto our skin. On the surface of the skin, the water evaporates and cools the skin. In those red tubes, blood flows.

All the things we encounter here are formed either by a whole lot of cells or by one single cell, such as the cells that swim in our blood.

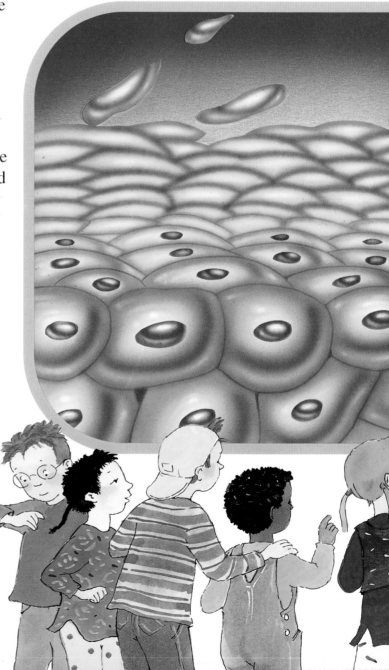

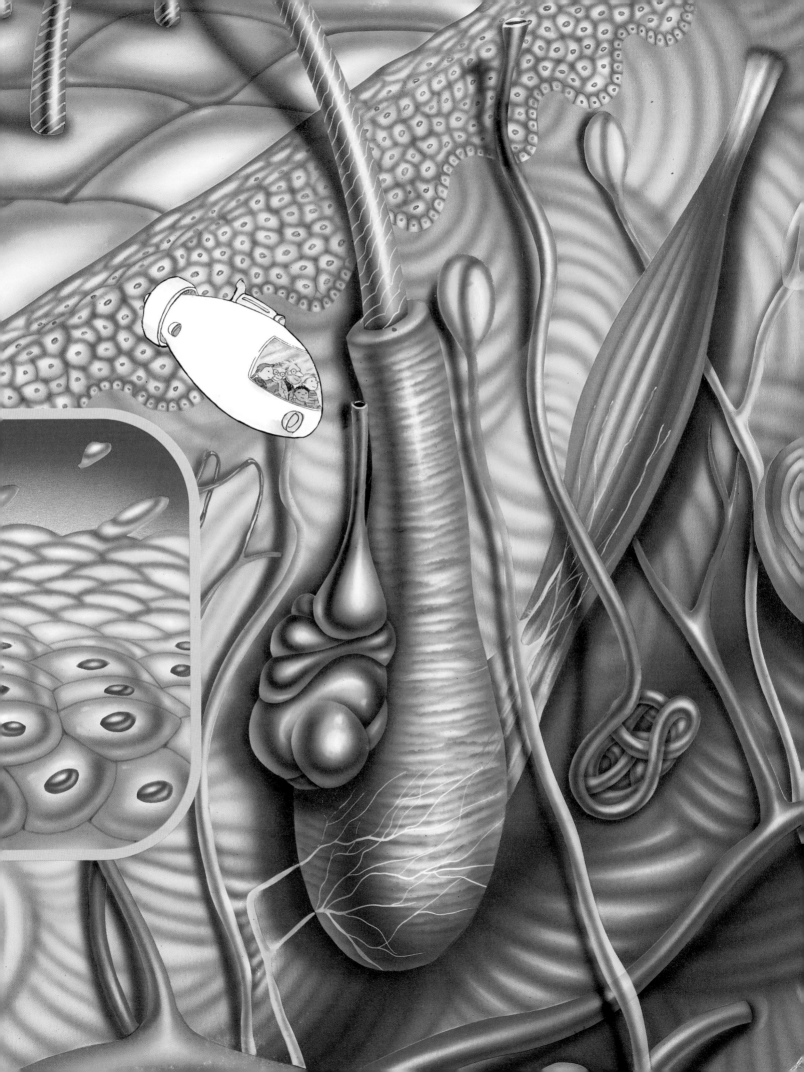

Cells are experts

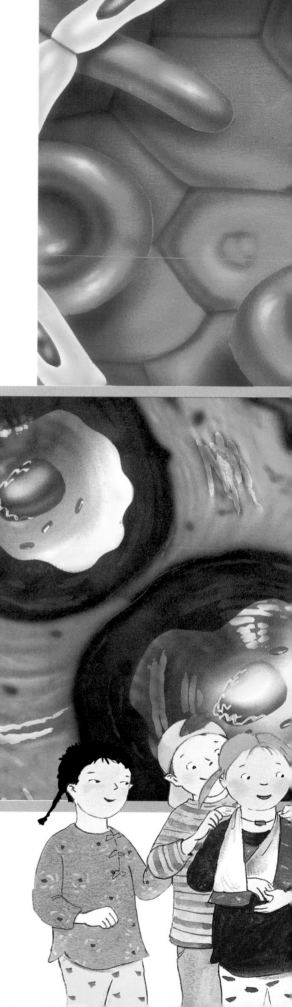

Hundreds of different kinds of cells form the human body and help to keep it alive. They have different shapes and different tasks to perform. In order to see them better, we need to go to Shrink Stage Two. Shhhhrink! Now we are among the individual cells. Most kinds of cells cling tightly together. Here, for example, we see endothelial cells that form the living building blocks for the walls of the blood vessels.

Gene, do all cells cling to each other?

Most kinds of cells do. They cling together tightly. Here, for example, we see endothelial cells, which form the living building blocks for the walls of the blood vessels. However, in the blood, some kinds of cells float quite alone. The rubber boat-shaped cells are the red blood cells. They have all lost their nuclei. Still, each one of them is an independent cell. They are loaded with oxygen, which they carry to all parts of the body. The white blood cells have a very different function. They are the patrolers that fend off alien invaders such as bacteria or viruses.

What happens to an old cell?

That depends on what kind of cell it is. Most cells do the same thing their whole lives. When they are old and used up, they die and are broken down into their components by cells called macrophages, *big eaters*. The remaining cells divide and replace the dead ones.

Other cells still fulfill useful tasks even after they are dead. See the cells that build up the hair root? At first they are very much alive. Then, they have to give up their lives, and the remaining parts are used to form hair. All that remains of them is a protein called keratin.

What are fingernails made of, Gene?

Fingernails are made of the same material as hair—keratin. The keratin in nails, however, is packed together much more densely than the keratin in skin cells. The deeper these cells are within the body, the more alive they are. With time, they wander toward the outer surface. When they reach it, nothing is left but a network of keratin fibers.

Are bone cells dead, too?

No, amazingly enough, they are fairly lively. Special bone cells constantly build up and break down the bone mass, which is made of minerals and proteins. That is why a broken finger bone can heal within a few weeks. The innermost part of the bone, the marrow, is in fact the home of cells that continuously divide to build fresh red and white blood cells.

Real teamwork

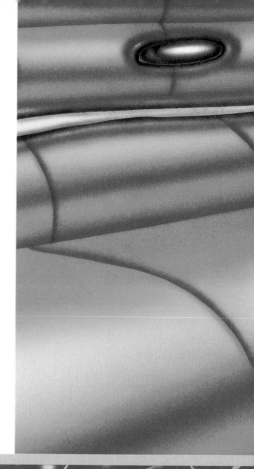

Cells are specialists that need one another. The finger can function only when all the different types of cells work together. The different cells that make up the finger have to exchange messages with other cells within the organism at certain times. Most of this work is done automatically. We don't notice much of it—except in some cases when the nerve cells are at work.

What does a nerve cell look like, Gene?

The nerve cells have widely stretched arms that connect with other nerve cells. Nerves are made of bundles of many such nerve cells. Some kinds of nerve cells are very long. The nerve cells that extend from the spine to the big toe can even be three feet long!

Why do we feel only the nerves?

Nerve cells are designed to transport signals. They reach almost every part of our body. They let us feel pain when a finger gets pinched. In our brains, billions of tiny nerve cells cooperate and deal with the signals in a way that makes us see, think, smell, and taste—or even lets us have thoughts and feelings. As a matter of fact, we can only discuss nerve cells because we have so many of them!

Nerve cells, by the way, are a good example of the way different kinds of cells depend on each other. What could a bundle of muscle cells do without proper connections to nerve cells? Nothing, really. Muscles respond to proper commands coming from nerve cells. Some signals are automatically sent to the right address, for example, the nerve signals that let the heart beat. Other signals have to be sent by choice, such as to make a finger move. The nerve center in the brain sends out a signal that arrives in the finger and tells the muscle to contract.

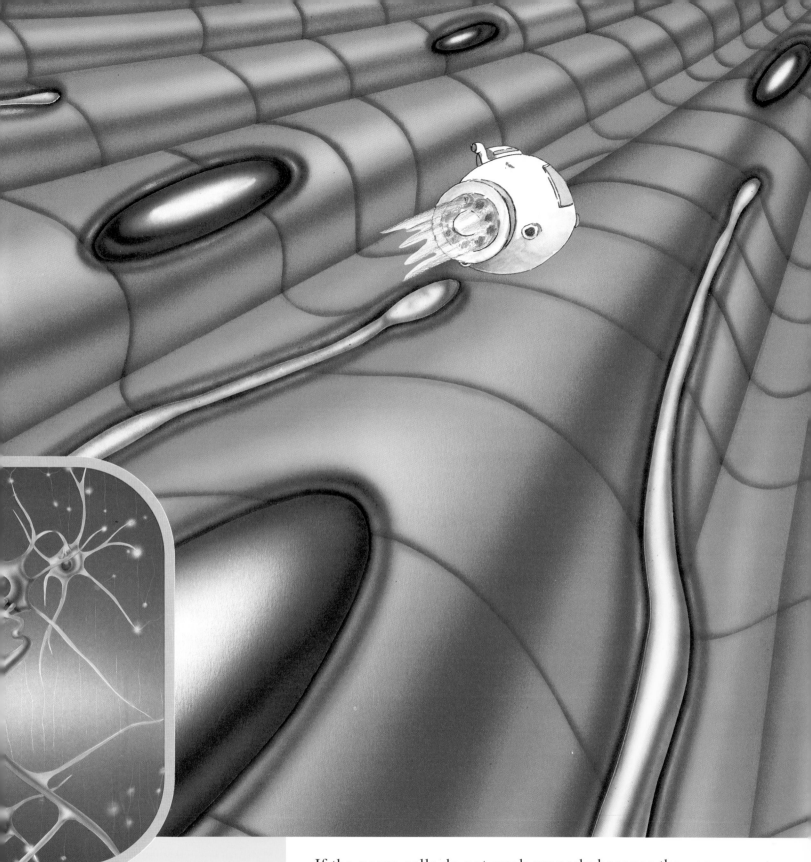

In our brain, billions of nerve cells are connected to each other. Here we see just a few of them. And here nerve cells connect to bundles of muscle cells in the finger.

If the nerve cells do not work properly because they are inflamed or even cut apart, the muscles cannot bend. That is the reason why, for example, a paralyzed leg cannot move.

Can cells do *magic ?*

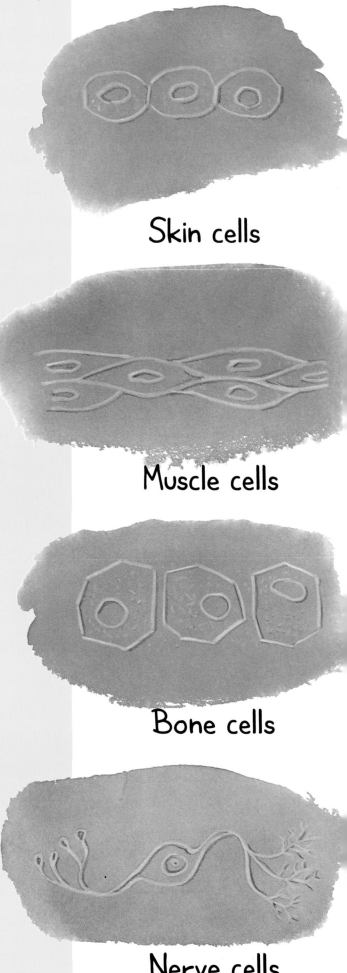

Skin cells

Muscle cells

Bone cells

Nerve cells

All living beings—humans, elephants, trees, mushrooms, and bacteria—consist of nothing but cells and of things produced by cells. Here are some typical shapes for you to compare.

Yet, in spite of their differences, all kinds of cells have much in common. Cells create fresh cells. They can move by themselves. They can recognize each other, can stick to each other, or—if need be—can fight each other. They send out and receive messages. By sticking together, talking to each other, and producing many kinds of substances, cells build a complicated organism like a human being, an elephant, a tree, or a mushroom. Now, why is it that cells can do all these things? Why is it that nonliving things like stones or grains of sand do not have these abilities?

Is something magical in the cells?

Certainly not. All cells consist of exactly the same kind of substances that nonliving things are made of—endlessly tiny balls called atoms. There are many kinds of atoms—solids like carbon, sulfur, and phosphorus; gases like oxygen, hydrogen, and nitrogen; and many, many others, such as metals.

Within nonliving things, the atoms are arranged in quite a simple and dull way. Therefore, a stone is just a stone. It can neither move by itself nor duplicate, grow, or recognize other stones. Within living things, however, the same or other kinds of atoms make extremely complicated structures. For example, atoms of carbon, sulfur, oxygen, hydrogen, and nitrogen can firmly connect to larger and more complicated molecules, such as sugars, fats, nucleotides, and amino acids.

Plant cells

Bacteria

We could call these organic molecules biomolecules because *bio* means *life*. Yet, they are still nonliving substances until they are put together within a cell in the right way and combine to form proteins, DNA, or membranes. Only then can atoms and molecules work together in a way that we call *life*. So the secret of life has very much to do with the way in which certain substances, such as proteins, are assembled.

Why assemble proteins, Gene? Don't we get them from food?

Yes, we eat proteins formed by the plant and animal cells in our food. When we digest these proteins, they are broken down into small units called amino acids. Our cells can rearrange those amino acids to build our own proteins. They are the most important molecules needed by cells. Certain proteins are the building blocks for cells. Other proteins are the tools that the cells use for their many kinds of activities.

Hence, cells make proteins. Proteins make cells. Many cells build an organism—a living creature.

Endoplasmic reticulum

Mitochondrion

Golgi apparatus

Nucleus

Cell skeleton

Ribosomes

Lysosomes

Centriole

The cell *works !*

Life is no easy matter—not even for a cell. Staying alive requires factories, power stations, transportation tracks, storage containers, and, most importantly, the right construction plan. We will now take a look at such a plan in a hair root cell. It could be any cell, though, for most of them consist of the same parts.

Each cell is covered by a very thin **membrane**, which keeps the cell shielded from the outside. The membrane is a greasy film composed of many different proteins that have different tasks to fulfill. They can receive from or send signals to other cells and make cells stick together. They choose what substances can enter the cell.

The parts floating inside a cell are called organelles. Each part is covered by its own membrane and moves around in a jelly-like soup called **cytoplasm**. The cell is given its shape with the help of elastic protein fibers, called the **cell skeleton**.

Within the **mitochondria**, the cell produces energy. **Lysosomes** shred, dispose, or store waste.

Ribosomes are tiny factories producing fresh proteins. Ribosomes either float freely in the cytoplasm or are bound to the **endoplasmic reticulum**. This is the place where proteins are made that are needed in the membrane or outside the cell.

Centrioles are in charge of organizing the cell skeleton when the cell divides and two new cells arise.

The **Golgi apparatus** adds sugar to fresh proteins and makes sure they reach their final destination.

The big ball in the middle of the cell is called the **nucleus**. This is where all the building plans for making fresh proteins are stored and copied into a message the ribosomes can read. Inside the nucleus, there are one or more smaller balls, each called a **nucleolus**, which assemble the various parts of the ribosomes.

All this—the shape, the function, and the cooperation of the organelles—relies on many, many special proteins that are so small we can't even see them here.

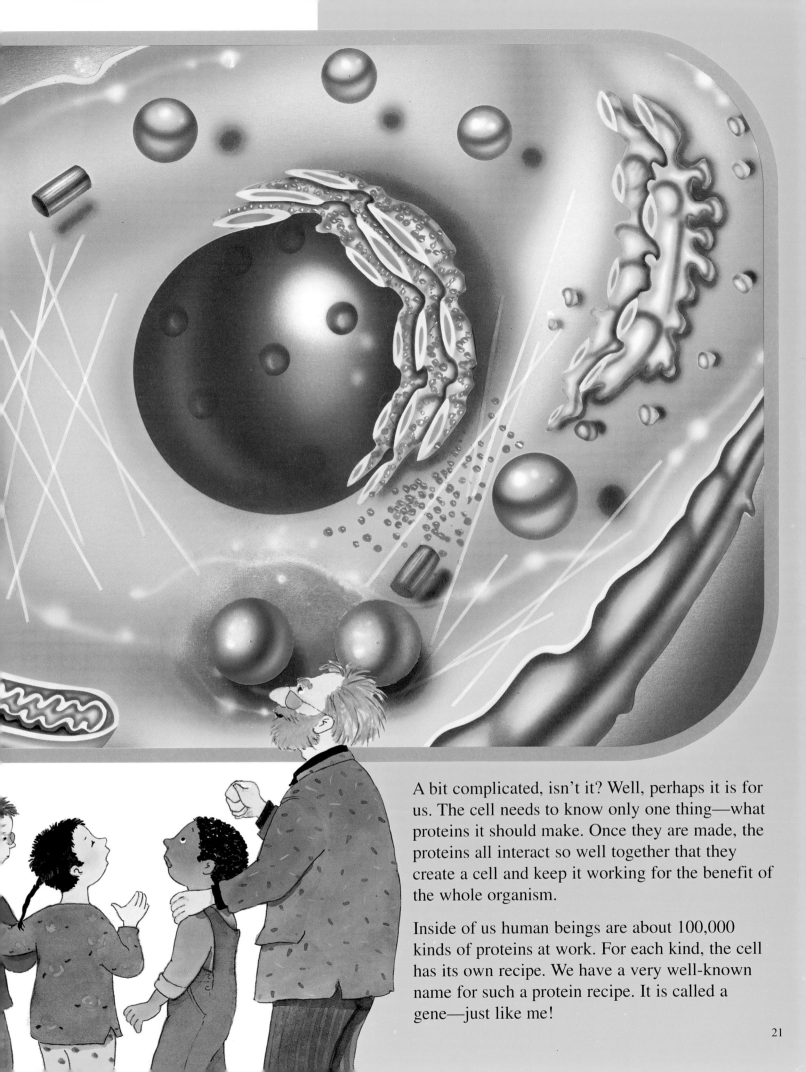

A bit complicated, isn't it? Well, perhaps it is for us. The cell needs to know only one thing—what proteins it should make. Once they are made, the proteins all interact so well together that they create a cell and keep it working for the benefit of the whole organism.

Inside of us human beings are about 100,000 kinds of proteins at work. For each kind, the cell has its own recipe. We have a very well-known name for such a protein recipe. It is called a gene—just like me!

Perfect planning

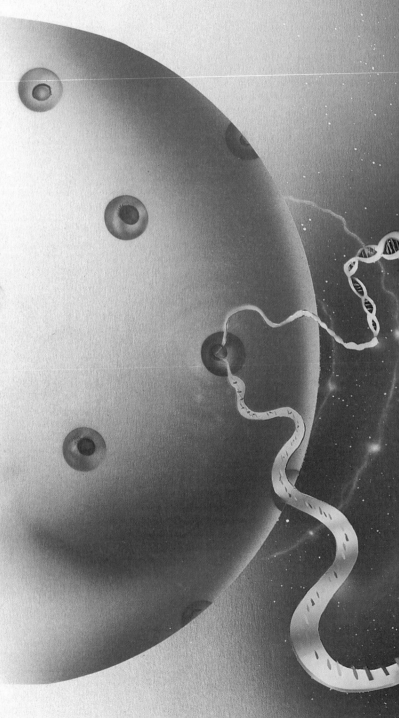

Now we are inside the cell. We have shrunk down so much that 100,000 MicroMachines would fit into the head of a pin! The blue ball we just passed is the cell's nucleus, in which all the genes are stored on unbelievably long, tiny, curiously twisted strands. They are called DNA, which is short for deoxyribonucleic acid. In humans, the entire DNA is divided into 46 pieces that we call chromosomes.

When unrolled, DNA has the shape of two threads that spiral around each other, connected by rungs—like a ladder. This is called a double helix. The rungs of the DNA ladder are made up of molecules called nucleotides. They come in four different types: adenine, thymine, cytosine, and guanine—abbreviated as A, T, C, and G, and shown here in different colors. Each rung is made up of a pair. A red A only connects with a blue T, and a yellow C only with a green G. They are, so to speak, the letters with which genes are written.

What exactly do genes do, Gene?

They make sure that our cells make the right proteins by putting together the right building blocks (called amino acids) to build those proteins. Each kind of protein is governed by a certain part of the DNA. This section is called a gene. On the gene responsible for keratin, for example, are several thousand code words. The number and order of the code words make the keratin gene different from every other gene and the keratin protein different from every other protein.

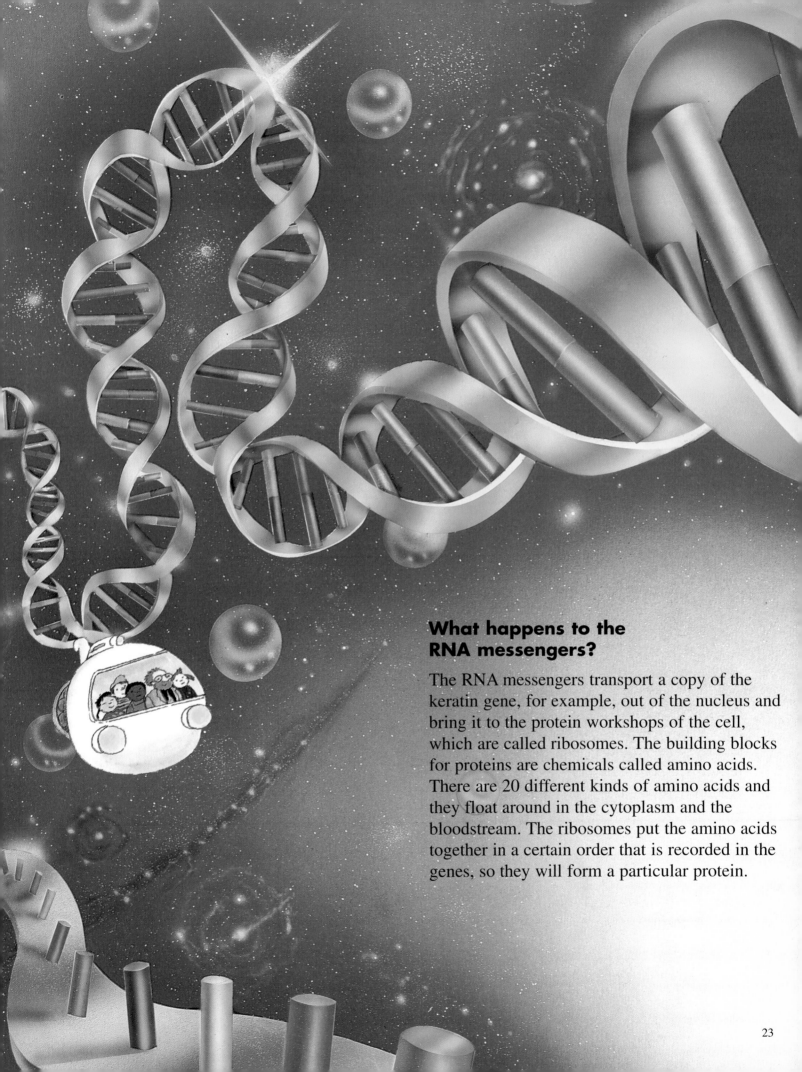

What happens to the RNA messengers?

The RNA messengers transport a copy of the keratin gene, for example, out of the nucleus and bring it to the protein workshops of the cell, which are called ribosomes. The building blocks for proteins are chemicals called amino acids. There are 20 different kinds of amino acids and they float around in the cytoplasm and the bloodstream. The ribosomes put the amino acids together in a certain order that is recorded in the genes, so they will form a particular protein.

How genes can make *proteins*

The genes are the step-by-step recipe for building the amino acids of proteins. These recipes are stored on the DNA, just like music is stored on magnetic cassette tapes. A tape without a player is of no use; the genes also need a kind of player. These players are the cells with their ribosomes.

All cells in our body have exactly the same DNA, which stores all 100,000 genes. Every cell could produce 100,000 different proteins, although it doesn't do so, of course. Only a few thousand types of proteins are actually formed by all types of cells. We call them household proteins because they are important for the basic functions. In addition to household proteins, each cell type makes other, specific proteins. Liver cells need other kinds of proteins than, for example, the skin cells. The skin cells need different kinds than the blood. In a particular kind of cell, there will be only as many genes in operation as there are proteins at work. These are the proteins that allow the cell to fulfill its own specific tasks and its household tasks.

How do genes get turned on, Gene?

This is done by special proteins. They attach themselves to the DNA near a certain gene. Beginning at a special spot, they make the two strands separate like a zipper. A portion of one strand of the DNA ladder is then used as a template to make a new thread of RNA, which is an exact copy of the gene. The RNA copy, called messenger RNA, is then set free, and the DNA ladder is zipped up again. The building blocks for the RNA copy of the gene are nucleotides that look very much like the DNA nucleotides. As soon as the gene has been completely copied, the RNA messenger is transported out of the nucleus. Turning on a gene simply means making a copy of it.

How the genetic code works:

On DNA, nucleotide pairs are lined up in a specific sequence to make genes.

The DNA unzips, and RNA nucleotides attach to form a strand with the complementary sequence. However, RNA does not contain thymine (blue T). Instead, the nucleotide uracil (violet U) is inserted. This copy of the gene is called messenger RNA. It carries its message to the ribosomes, where proteins are manufactured.

Three nucleotide letters form a code word. That word specifies one of the 20 different amino acids that build proteins. The sequence of the nucleotide words determines the exact sequence to align the particular amino acids. Once the right amino acids are put together in the right order, they curl up to form a specific protein.

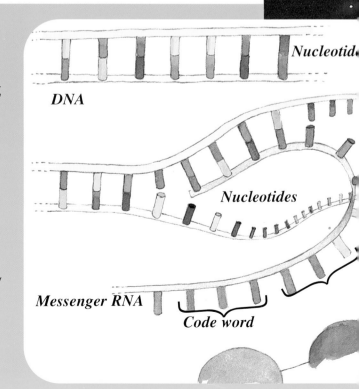

Nucleotide

DNA

Nucleotides

Messenger RNA

Code word

What happens with the copy of the gene?

The messenger RNA connects to a ribosome. This is the real workshop for proteins. The copy of the gene is now translated—using the gene recipe—into a chain of amino acids. The protein building blocks floating in cytoplasm, the amino acids, are strung together step-by-step. Finally, a new protein string emerges from the ribosome that immediately folds up into a ball or a fiber.

Most proteins are needed to build or to work in the very cell in which they are created. Each cell makes its own parts all by itself—for example, keratin. Other proteins have to fulfill their tasks in the membrane of the cell, in an organelle, or outside of the cell—for example, in the blood.

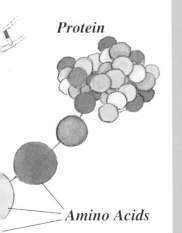

Protein

Amino Acids

Watch out!
Skeletons!

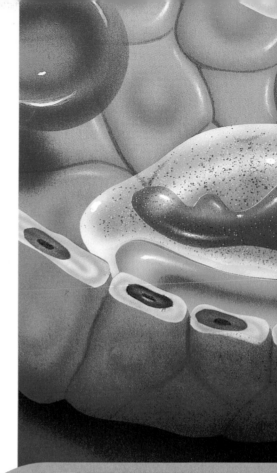

Our body needs about 100,000 different kinds of proteins to work properly. The DNA with the genes provides the plans to build proteins. However, the proteins carry out the plans. They are the tools for building a cell and an organism. They are the craftspeople that do the work. They provide most of the building blocks. All proteins have very different characteristics and can do very different jobs. For example, some proteins linked with DNA make sure that the correct gene is copied. Other proteins, such as keratin, support the cell and form its skeleton.

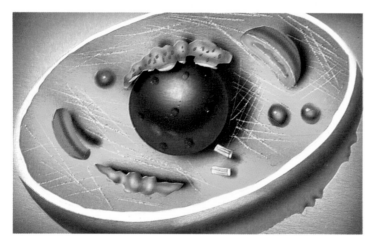

Why do cells need a skeleton?

The skeleton of the cell is made up of thin fibers. It has many tasks. When the cell divides, it makes sure that each of the daughter cells receives the same number of chromosomes. The skeleton makes it possible for the cells and their organelles to move about. It also maintains the shape of the cell.

Let us look, for example, at these cells that coat the inside of the intestines. They have supporting fibers, which are made of the protein tubulin. Just like the poles of a tent, the tubes of tubulin support the membranes of the intestinal wall cells. The membranes fold like brushes and achieve a larger surface. This is the best way to bring these strangely shaped cells into contact with the many nutrients passing through the intestines.

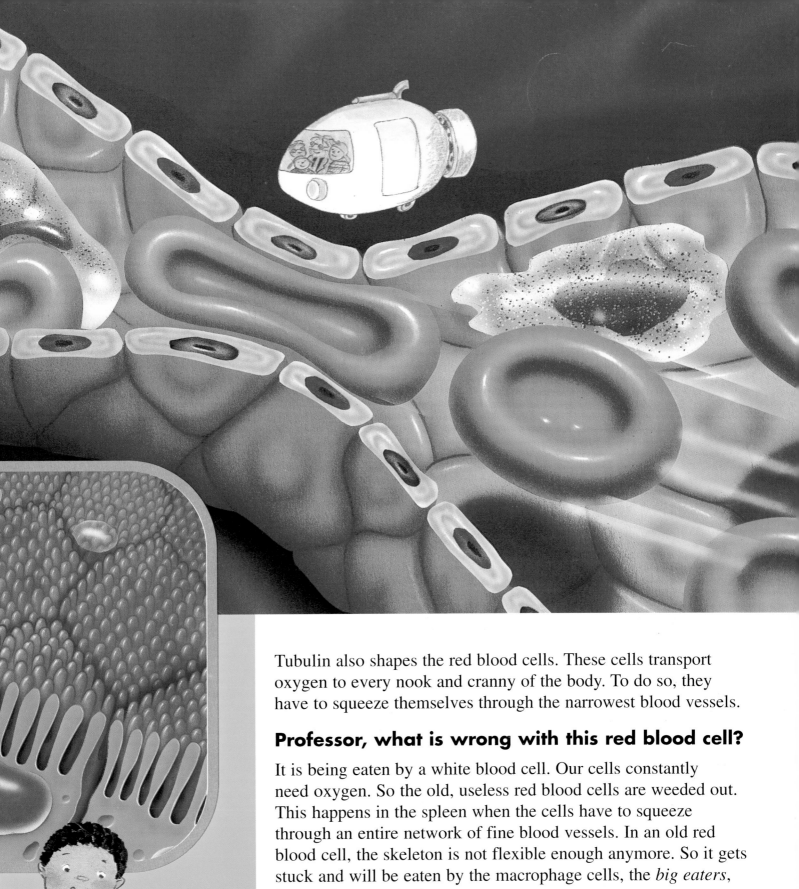

Tubulin also shapes the red blood cells. These cells transport oxygen to every nook and cranny of the body. To do so, they have to squeeze themselves through the narrowest blood vessels.

Professor, what is wrong with this red blood cell?

It is being eaten by a white blood cell. Our cells constantly need oxygen. So the old, useless red blood cells are weeded out. This happens in the spleen when the cells have to squeeze through an entire network of fine blood vessels. In an old red blood cell, the skeleton is not flexible enough anymore. So it gets stuck and will be eaten by the macrophage cells, the *big eaters*, just as the name indicates.

Other fiber proteins of the cell skeleton, namely actin and myosin, have different tasks. They can glide along each other's surfaces and move things from one place in the cell to another—or even move the whole cell. We'll see how that works in just a minute.

In perpetual *motion*

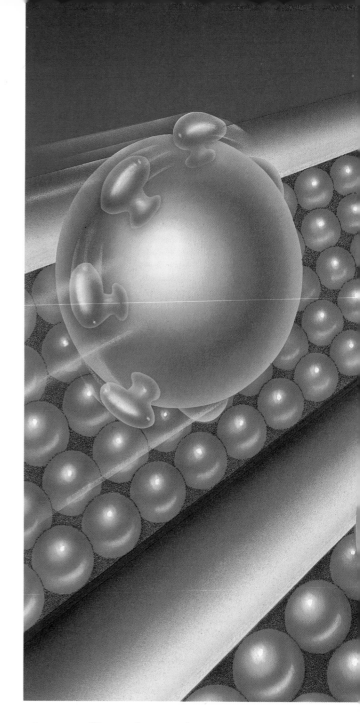

All our cells are in constant motion. Usually, we notice only when our muscle cells move. Form a fist. The muscles in the lower arm that you can feel get tense only because several million muscle cells are pulling together at the same time. Countless cells have to move at once in order for us to blink or crook a finger or for the heart to pump blood through the body.

Professor, how do the muscles pull together?

Well, let's just take a look. Muscles consist of long bundles of muscle cells. In the cytoplasm, the fibers of actin and myosin are arranged alongside each other in a very ingenious manner. One end of the actin fiber and one end of the myosin fiber are firmly attached to disks, while the other end of each one is loose—although they can connect with each other. When the muscle cell receives the right signal, the myosin fibers run along the tiny strands of the actin faster, making the whole cell shorten. When millions of cells do this at the same time, the muscle is strong enough to let us pick up a heavy suitcase.

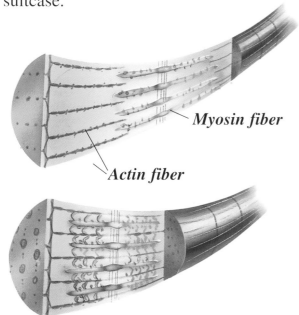

Myosin fiber

Actin fiber

Can cells take a journey?

Yes, of course. Some of them just drift along in the blood, for example, the red blood cells. Sometimes, the white blood cells also go off on their own. When an inflammation develops somewhere in the body, the cells that form the blood vessels become sticky—at least to the white blood cells. Do you see how they are gathering here? Now they squeeze between the blood vessel cells and leave the bloodstream. They can then reach the inflammation site outside of the blood vessels and fight the germs.

What is that running around inside the cells?

There is also a lot of movement inside the cells. Countless little particles are always being created, which have to get to the right spot at the right time. Proteins destined for the outer part of the cell are packed into membrane bubbles and sent off. These little bubbles are called vesicles. They can run on little tubes called microtubuli, which are made of the protein tubulin. Vesicles have proteins that run like little feet along the microtubuli. Usually, they travel only a fraction of a millimeter. In most nerve cells, however, the vesicles have to travel long stretches from the nerve cell body to the outer extremities—back and forth, even up to three feet.

Food for our cells

Cells have many things to do and therefore need lots of energy. The fuel they constantly need to be fed is brought by the blood. In order for cells to be fed, we need to eat. Of course, a cell doesn't know what to do with a cookie. The cookie first must be broken down into smaller parts. This takes place during digestion when the useful parts of a cookie, such as sugar, fat, and amino acids, pass through the walls of the intestines into the blood to end up as the energy supply in our cells. All three kinds of nutrients are finally burned in the mitochondria—the power plants of the cells.

The cell fire in the mitochondria is not especially hot—just warm enough to keep the body warm. Naturally, there are no blazing flames. Still, the same things happen in the cell's mitochondria as in a real fire. Certain materials join with the gas oxygen, and energy is set free. This energy is used by the cells for various purposes in different locations. So the energy that is won through the burning process must first be stored. The best way would be in batteries.

Are there batteries in the cells?

Yes, at least there is something similar. In the cells, tiny molecules are floating around that can be considered chemical batteries. Here is a drawing of one. The cell fire in

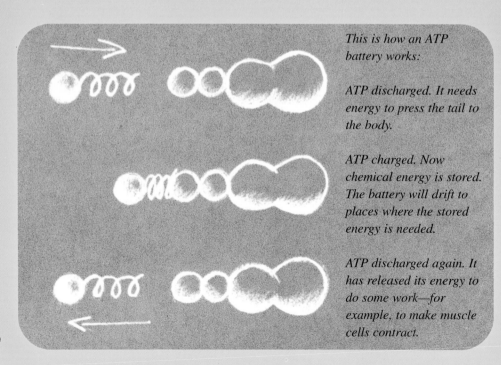

This is how an ATP battery works:

ATP discharged. It needs energy to press the tail to the body.

ATP charged. Now chemical energy is stored. The battery will drift to places where the stored energy is needed.

ATP discharged again. It has released its energy to do some work—for example, to make muscle cells contract.

the mitochondria loads them by adding a phosphate particle. The loaded battery molecule, which we call ATP, can move to all nooks and crannies of the cell. Where energy is needed, certain proteins are waiting to uncover the phosphate particles. This sets free the chemical energy the cell uses to do almost everything it needs to do. The cell transports all kinds of things, creates warmth, divides itself, and moves. In this way, it serves the organism to which it belongs.

Do you remember how muscle cells work? Well many thousand ATPs allow the strands of the myosin fibers to move along the actin fibers.

Clever *membranes*

In membranes that surround the cells and their organelles, many different proteins are embedded that have many different tasks to fulfill. They act as doorkeepers, as receivers for messages, or simply as glue that holds their companions together.

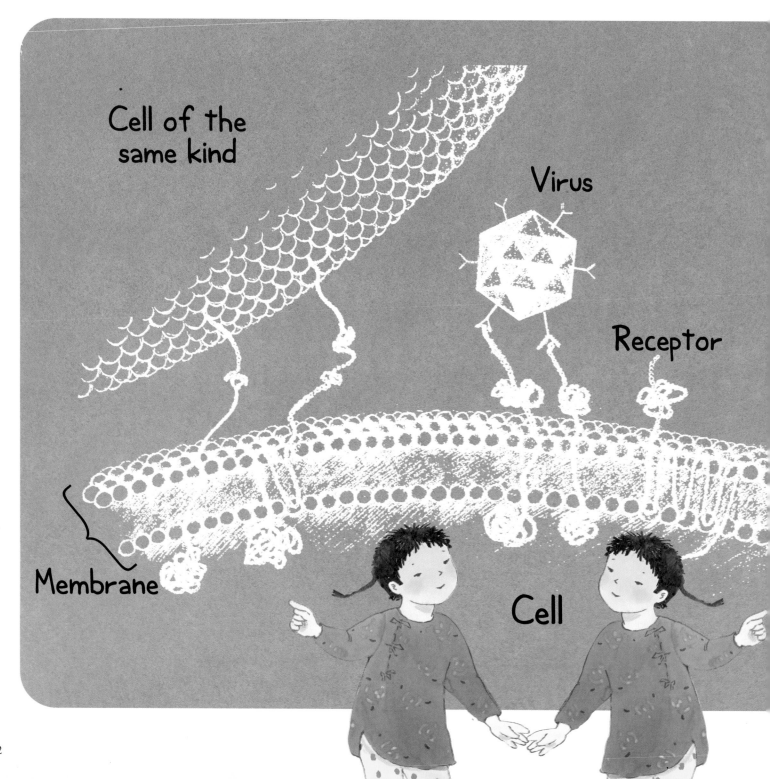

Cell of the same kind

Virus

Receptor

Membrane

Cell

What is allowed to enter a cell?

Fats, sugars, and amino acids are welcome in the cell; they are used to create energy or to build proteins and other molecules. Proteins found in the blood and other waste products must stay outside. The doorkeeper proteins keep track of this by acting as a canal in which the shapes of the oncoming molecules are identified. One could say that the doorkeepers check the chemical passports of the newcomers. They keep certain things out of and let other things into the cell.

The proteins that recognize the messages and receive them are called receptors. These receptor proteins also move like buoys in the membrane but have no entryway canals. Thanks to their special form, they remember the proper signal molecules. They react to a remembered signal and initiate the cell's answer to it. Such signal molecules are called hormones.

How does a hormone work, Professor?

There are many kinds of hormones that work in different manners. For example, if we are excited or afraid, a gland located on the kidney pours a large amount of a hormone named adrenaline into the blood. The adrenaline flows by the cells. In the membranes of the muscle cells that surround the blood vessels are adrenaline receptors. The receptors catch and then react to the message and send a signal to the muscle cells, telling them to pull together and tighten the blood vessels. As a result, the heart must pump faster and stronger. The blood brings more oxygen and sugar to the cells, which quickly— and only for a short time—can create more energy. At that moment, we become either alert and (we hope) strong enough to deal with the dangers at hand, or fast enough to escape.

How do cells stick to one another?

Other proteins in the cell membrane function as if they were intelligent glue. They join together with proteins that sit in the membranes of the neighboring cells and glue together cells of the same type. The cells of the intestinal walls, for example, are so thick and tightly joined that they hold the pulpy nutrients and bacteria within the intestines. Only digested nutrients may enter intestinal cells. In a similar way, the cells of the blood vessels, the skin, or the muscles form a strong and solid web or tissue by uniting the cells together.

Hormones

Protein

Friend or *foe?*

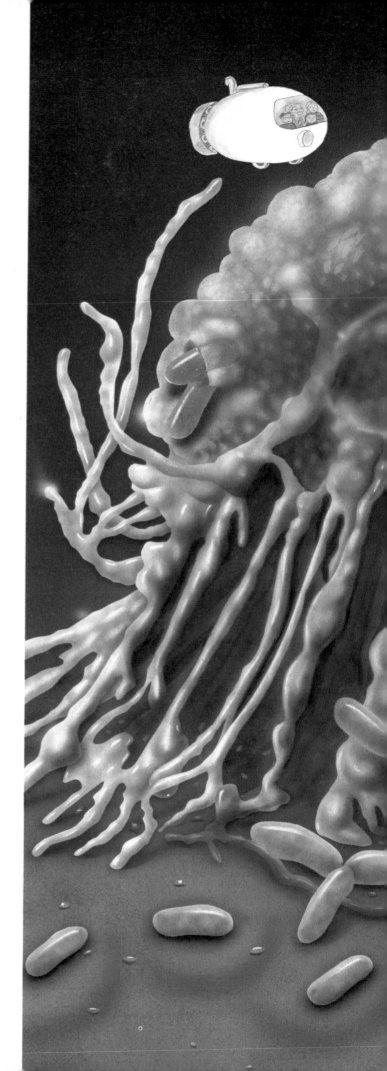

Our cells always want to stick to their own kind. In this way they prevent foreign cells, such as bacteria or malaria parasites, from getting into the blood, where they could multiply at an amazing speed. Bacteria are smaller than body cells, and they come in thousands of types. Some kinds kill the body cells by poison or destroy them with their digestive proteins.

Are all bacteria dangerous?

Fortunately, the answer is no. On the skin of each one of us are more bacteria than people on this earth. On the right we can see such a colony. Other bacteria live quite harmlessly in the intestines, where they are even useful. They can help, for instance, to digest the food we have eaten.

Many germs enter the body when we eat and breathe. Others enter through a wound when the bloodstream is open and accessible. To fight against these invaders, we have a whole army of defense cells, the white blood cells. They are formed in the bone marrow for the very purpose of defeating invaders. Here we can see how a defender cell catches green bacteria with its tentacles and devours them.

Some kinds of white blood cells specialize in fighting very specific kinds of germs. One group, the B cells, creates weapons called antibodies. Antibodies attach themselves to the invaders, paralyzing and marking them as enemies. Another type of white blood cells, the killer T cells, track down body cells in which a virus has lodged itself. They destroy the cell along with the virus. Killer T cells also destroy cancer cells, those cells that have forgotten to stop dividing.

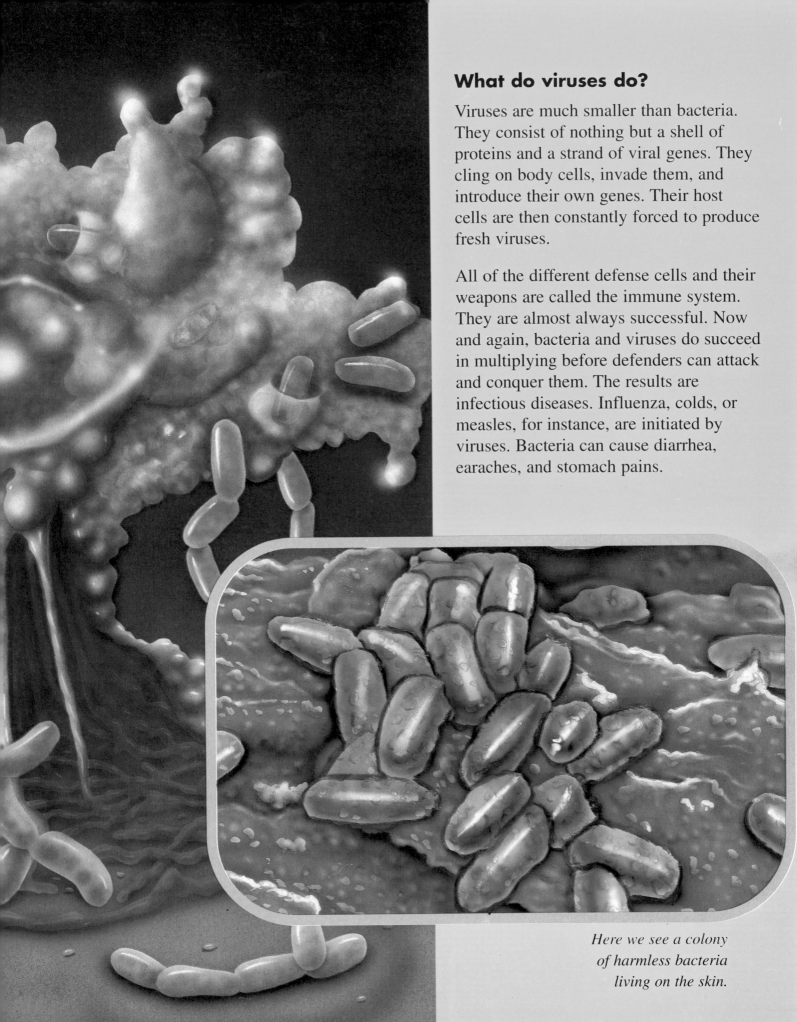

What do viruses do?

Viruses are much smaller than bacteria. They consist of nothing but a shell of proteins and a strand of viral genes. They cling on body cells, invade them, and introduce their own genes. Their host cells are then constantly forced to produce fresh viruses.

All of the different defense cells and their weapons are called the immune system. They are almost always successful. Now and again, bacteria and viruses do succeed in multiplying before defenders can attack and conquer them. The results are infectious diseases. Influenza, colds, or measles, for instance, are initiated by viruses. Bacteria can cause diarrhea, earaches, and stomach pains.

Here we see a colony of harmless bacteria living on the skin.

35

Where do *cells* come from?

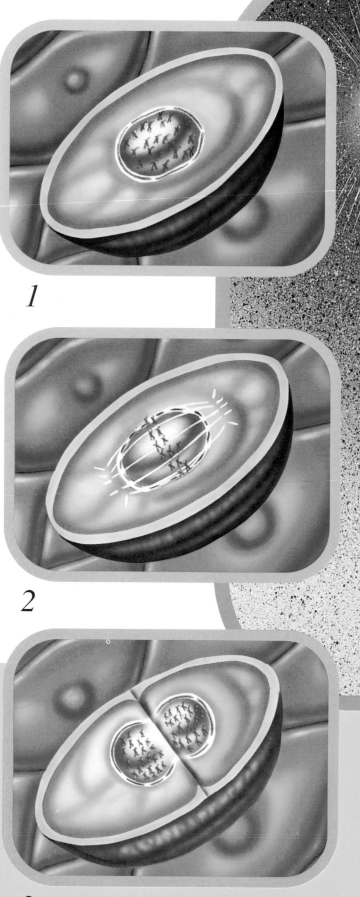

1

2

3

Whenever fresh cells are needed, a parent cell divides itself into two parts. Two new cells are formed, called daughter cells. What a great idea! Parents turn into children again!

Where does the parent cell come from?

All of the different cells that we consist of are descendants of one single cell—a fertilized egg that once was in the body of our mother. At the time of conception, about nine months before we were born, a sperm cell from our father mated with the egg cell. At that time, the genes from the father cell and the mother cell mixed together. Thus, each of us began from a single cell that received the genes of both mother and father. The egg immediately began to divide over and over again until, gradually, all the different types of cells that a human organism needs were developed.

When does a cell make new cells?

Genes in a cell's nucleus control cell division. They also tell the cell when to stop dividing. Sometimes accidents happen and sick cells that have damaged control genes keep on dividing needlessly. This can result in a tumor, a large, malignant lump of cells, which is called cancer.

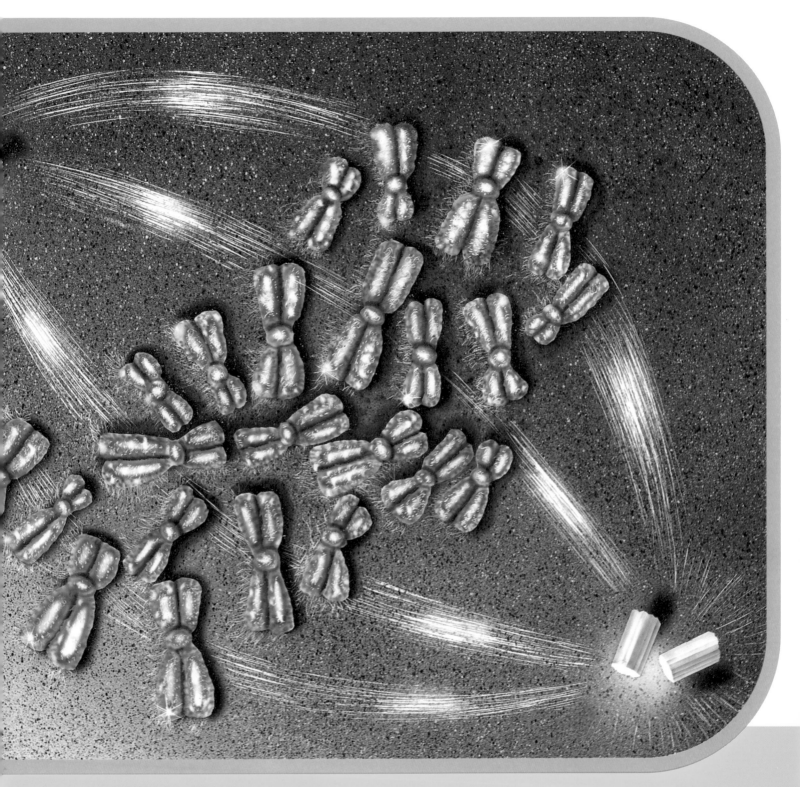

1) *When a cell divides into two, the DNA strands double. Now they appear as shapes called chromosomes. Each cell has 23 pairs of chromosomes.*

2) *The chromosomes line up properly.*

3) *Both new cells, called daughter cells, receive one complete copy of the DNA after the chromosomes are equally distributed to the new daughter cells.*

Cell division is not so easy. The new daughter cells must be more or less identical to the parent cell. They need the same genes so that they can create the same proteins. Therefore, equal distribution of the DNA is key to cell division. The DNA wrapped up in the chromosomes must be doubled; each daughter cell can then receive its own copy of chromosomes. Here we can see how that works. Pay attention to the centrioles—they put the microtubli in the right order. These are the routing tracks that ensure that the chromosomes are distributed equally to the new cells.

Goodbye,
cells!

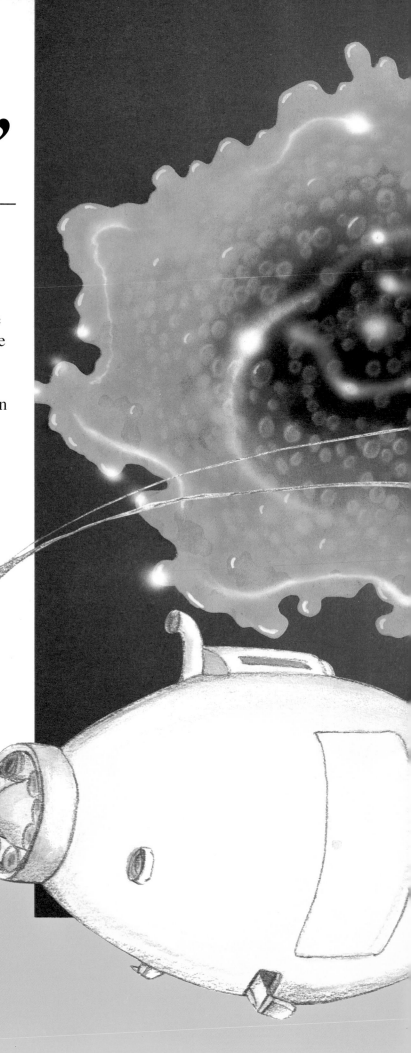

Different kinds of cells divide at different intervals of time. For example, the cells lining the intestines are used up after just three days and then need to be replaced. Our nerve cells, on the other hand, stop dividing when we are still babies. We have to live with these same cells for the rest of our lives. Our brain cells, which can be destroyed by drugs, cannot be replaced either. So take good care of them! Skin cells, in turn, are in a better situation. After three weeks, they are replaced with fresh cells.

Here we are again, back at the starting point of our expedition. Since we entered the MicroMachine, thousands of skin cells have died. When we rub our hands now, what is the result? Sweat, dust, and dead skin cells, along with remainders of all those chromosomes, proteins, and ATPs that we encountered during our journey, come off.

And now?

Enlaaaaaaarge!

What do you remember?

Proteins consist of what?

a) amino acids
b) nucleotides
c) DNA

a) is correct. There are 20 kinds of amino acids. They could be assembled to make millions of different proteins—just the same as 20 notes could make millions of tunes. A human organism, however, only makes about 100,000 different proteins. Nucleotides are the *letters* that make the genes stored on DNA (deoxyribonucleic acid).

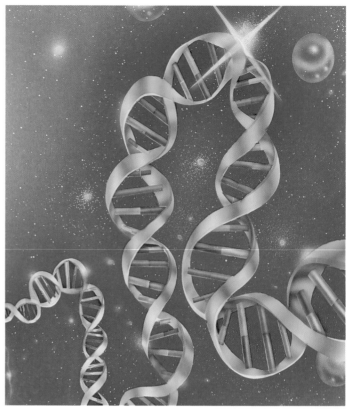

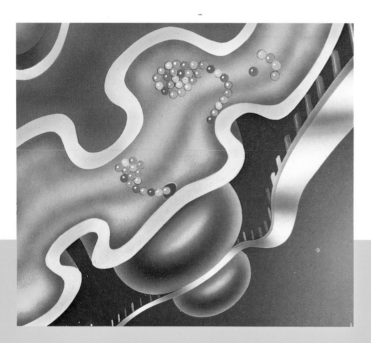

A gene is

a) a piece of DNA holding a message
b) the name of our tour guide
c) a recipe for making a protein

a) and c) are correct. Our genes contain information that determines, for example, our sex, the color of our skin, the length of our nose, and also which diseases we are likely to get. As for answer b, don't be silly!

The genetic code

a) is the whole library of protein recipes
b) takes a copy of the gene to the ribosomes
c) is the way gene letters build code words for amino acids

c) is correct. A block of three nucleotides on the DNA makes up the code word that stands for a specific amino acid. The messenger RNA then brings a copy of the gene to the ribosome. The library of genes within the cells of a living being is called its genome.

There are genes in

a) humans and animals only
b) humans, plants, and animals only
c) all living beings, including bacteria

c) is correct. The genome (the list of all genes) of a typical bacterium consists of about 3,000 genes. Humans have about 100,000 genes. Different genes make different beings. Snail genes differ quite a lot from human genes, while chimpanzee genes differ only slightly from ours.

Cells contain
a) power plants
b) protein plants
c) building plans for proteins
d) tiny architects drawing the plans

a), b), and c) are correct. Of course there are no tiny achitects within the cell. It has received its plans (or genes) from its parent cell.

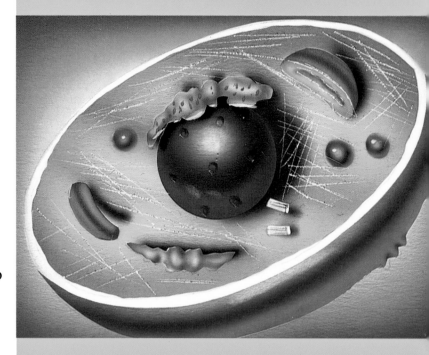

Chromosomes are
a) small animals inhabiting a cell
b) pieces of the DNA
c) the skeleton fibers of the cell

b) is correct. Within human cells, DNA is broken into 46 pieces called chromosomes. Cabbage has 18 chromosomes. The number of chromosomes doesn't seem to correspond with the intelligence of the being. Dogs have 78 and goldfish 94 chromosomes. There are no tiny animals living in a cell. The cell skeleton is called cytoskeleton.

Cells recognize each other
a) from the taste of the mitochondria
b) from the structure of proteins in its membrane
c) from the flavor of the cell soup

b) is correct. Each kind of protein has its specific structure, and specific proteins match like a key and keyhole. They sit in the cell membrane and cling to proteins sitting in neighboring cells of the same kind. This way, cells recognize which cells are of the same tissue type and which are not.

Which consists of only one single cell?
a) nerves
b) muscles
c) bacteria

c) is correct. Muscles and nerves are made from bundles containing many thousands of cells.

Cells multiply by

a) division
b) laying eggs
c) pollinating

a) is correct. A cell divides by splitting into two, and two cells become four, and so on. A single bacterium can multiply to become a million bacteria, provided there is enough food. Our body cells, however, normally divide only when daughter cells are needed for the sake of the organism. Cells do not lay eggs. However, a chicken egg is the chicken's egg cell. Only plants reproduce through pollination.

Children grow because

a) their cells enlarge
b) their cells multiply
c) they really want to

b) is correct. The cells of children and adults are the same size. Within children, they multiply faster—so the whole body becomes bigger and larger, whether you want it to or not!

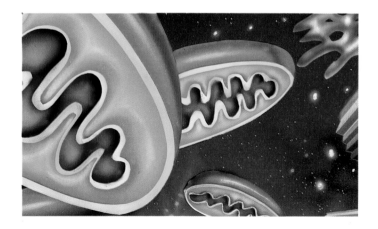

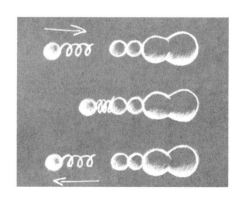

More fascinating facts about cells:

If a cell were as big as a house:

—a lysosome would be as small as a chair,
—a ribosome as small as a Ping-Pong ball,
—the nucleus as big as a room,
—the DNA as thin as a fishing line.

Nerve cells in our brain are our smallest cells. It would take 40 brain cells to cover the period at the end of this sentence.

It takes a cell one minute to load and unload two million ATP batteries.

About 1,000 mitochondria (power stations) and millions of ribosomes are within a liver cell.

A single blood drop contains about 200 million red blood cells.

Glossary

Actin and myosin proteins that work together to make muscle cells contract.

Amino acids the building blocks of proteins. There are 20 kinds of amino acids.

Antibodies proteins that help kill germs.

Atoms the smallest fragments of an element that can take part in a chemical reaction.

ATPs battery molecules that have been loaded in the mitochondria. They will set free their chemical energy wherever it is needed.

B cells white blood cells that make antibodies.

Bacteria foreign cells that are smaller than body cells. They come in thousands of kinds. Some are helpful to our body, some are dangerous.

Biomolecules complicated chemical substances such as sugars, fats, nucleotides, and amino acids. "Bio" means "life."

Bone cells busy in building up and breaking down the bone mass.

Brain the central control system of the body. It consists of billions of nerve cells. It receives and sends messages along a network of nerve fibers to nearly all parts of the body.

Cell division the way cells multiply. A parent cell divides itself into two parts. The fresh cells are called daughter cells.

Cell skeleton consists of various kinds of protein fibers.

Cells tiny living beings. Hundreds of different kinds of cells form the human body and help to keep it alive. The main task of a cell is to make proteins.

Centrioles organelles that help cells to get organized during cell division.

Chromosomes packages of DNA containing the genes. In the nucleus of almost every one of our cells are 46 chromosomes in 23 pairs.

Cytoplasm jelly-like soup in which the parts of a cell float.

DNA short for deoxyribonucleic acid. Genes are lined up on these incredibly thin strings, shaped like twisted ladders, that are found in each cell.

Endoplasmic reticulum the place where ribosomes make proteins that will do their job in the membrane or outside the cell.

Endothelial cells form the living building blocks for the walls of the blood vessels.

Genes sections on the DNA. Stored in the cell nucleus, they are the recipes for making proteins. Each gene consists of several thousand code words.

Golgi apparatus a cell organ (organelle) that helps send out proteins.

Hormones signal molecules that make cells work in a specific way. The hormone adrenaline, for example, tells muscle cells around the blood vessels to contract.

Human egg cell contains only 23 single chromosomes. In order to divide and form all the different cells an organism needs, a female egg cell needs to be fertilized by a male sperm cell bringing another set of 23 chromosomes.

Human sperm cell contains only 23 single chromosomes. Male sperm cells can fuse with a female egg cell in order to fertilize it.

Immune system fends off germs. It consists of billions of white blood cells and of the weapons they produce.

Keratin a protein that builds up hair and fingernails and the skeletons of cells.

Killer T cells white blood cells that destroy body cells infected by viruses.

Lysosomes organelles that shred, dispose of, or store waste.

Macrophages ("big eaters") a kind of white blood cell. They patrol with the blood and fight off microbes such as bacteria by eating them.

Marrow the innermost part of bones where fresh red and white blood cells are made.

Membranes moist films composed of many different proteins. The membranes keep the cells shielded from the outside, make cells stick together, and receive and send out signals.

Messenger RNA takes a copy of the gene (of the protein recipe) to the ribosomes where proteins are made.

Mitochondria the power stations of the cell.

Molecules the smallest amount of chemical substance that can exist alone.

Muscle cells can contract. They form bundles that make our body move.

Nerve cells connect with other nerve cells to make bundles called nerves. Nerve cells transport signals. For example, nerve cells in the skin allow us to feel pressure, heat, and cold. Many billions of nerve cells cooperate in our brain to make us see, hear, feel, and think.

Nucleotides molecules that build the rungs of the twisted DNA ladder.

Nucleus the ball in the middle of the cell in which the genes are stored.

Nutrients food for our cells. These carbohydrates, fats, proteins, minerals, and vitamins come in with the food we eat.

Organelles the parts floating inside a cell.

Organism a complete living thing. It can consist of only one cell, such as a bacterium, or of many billions of cells, such as plants, animals, and humans.

Proteins the building blocks of cells and the tools that the cells use for their many kinds of activities. Proteins make cells, and many cells build an organism. There are 100,000 different kinds of proteins working together in the human body.

Receptors proteins in the cell membrane that recognize outside chemical messages.

Red blood cells carry oxygen to all parts of the body.

Ribosomes tiny balls within a cell that make fresh proteins.

Skin cells form layers that cover the inside and outside of our bodies.

Tubulin a kind of protein that makes supporting fibers that keep cells, such as red blood cells, in shape.

Tumor a large lump of cells that can arise when faulty genes tell cells to divide much faster than other cells.

Vesicles tiny membrane bubbles in which are packed freshly made proteins.

Viruses very small germs that have no life on their own, but can change the way our body cells behave. This way they can cause diseases.

White blood cells come in many different kinds. They are constantly on the outlook for harmful germs that they destroy.

Index